SPOT

BEST EVER ANIMALS

ANIMAL BABYSITTERS

by Anastasia Suen

amicus
LEARNING

ostrich  gorilla

Look for these words and pictures as you read.

whale  meerkat

Animal mothers need to eat.
Who will watch their babies?
Other animals will babysit!

Gorilla moms go look for food.
One dad stays home.
He watches the little ones.

gorilla

Giraffe moms help each other.
One mom stays with the babies.

Meerkats take turns.
Some stay home with the pups.
They keep the babies safe.

meerkat

Ostrich babies stay in a group.
One mom or dad cares
for all of them.

ostrich

whale

Whale families swim
in a big group.
They help each other.
Dads swim with other babies.

Babysitters help out.
They are the best!

ostrich gorilla

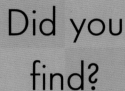

Did you find?

whale meerkat

Spot is published by Amicus Learning, an imprint of Amicus
P.O. Box 227, Mankato, MN 56002
www.amicuspublishing.us

Copyright © 2025 Amicus.
International copyright reserved in all countries.
No part of this book may be reproduced in any form
without written permission from the publisher.

Library of Congress Cataloging-in-Publication Data
Names: Suen, Anastasia, author.
Title: Animal babysitters / by Anastasia Suen.
Description: [Mankato, MN] : Amicus Learning, [2024] |
Series: Best ever animals | Audience: Ages 4–7 | Audience:
   Grades K–1 | Summary: "Animal families have babysitters
   too! Many species help take care of each others' young.
   This low-level search-and-find beginning reader reinforces
   new vocabulary with close-up images of animal families in
   their natural habitats. A great early STEM book to inspire
   learning about animals and life science for kindergartners
   and first graders"—Provided by publisher.
Identifiers: LCCN 2023038606 (print) | LCCN 2023038607
   (ebook) | ISBN 9781645492542 (library binding) |
   ISBN 9781681527789 (paperback) | ISBN
   9781645493426 (pdf)
Subjects: LCSH: Familial behavior in animals—Juvenile
   literature. | Parental behavior in animals—
   Juvenile literature.
Classification: LCC QL761.5 .S839 2024 (print) | LCC
   QL761.5 (ebook) | DDC 591.56/3—dc23/eng/
   20231204
LC record available at https://lccn.loc.gov/2023038606
LC ebook record available at https://lccn.loc.gov/
   2023038607

Printed in China

Rebecca Glaser, editor
Deb Miner, series designer
Emily Dietz, book designer
   and photo researcher

Photos by 123RF/mattiaath,
6-7, utopia88, cover; Alamy/
Tim Ireland, 1; Deposit Photos/
imagebrokermicrostock, 3; Getty
Images/Hao Jiang/500px, 14, Ibrahim
Suha Derbent, 4-5; Minden/Doug
Perrine, 12-13; Pixabay/polyfish, 10-11;
Shutterstock/Gaston Piccinetti, 8-9

ANIMAL BABYSITTERS